Henry Onderdonk

An Historical Sketch of Ancient Agriculture, Stock Breeding

and Manufactures in Hempstead

Vol. 1

Henry Onderdonk

An Historical Sketch of Ancient Agriculture, Stock Breeding and Manufactures in Hempstead
Vol. 1

ISBN/EAN: 9783337145835

Printed in Europe, USA, Canada, Australia, Japan

Cover: Foto ©berggeist007 / pixelio.de

More available books at **www.hansebooks.com**

AN

HISTORICAL SKETCH,

OF

ANCIENT AGRICULTURE,

STOCK BREEDING AND MANUFACTURES,

IN

HEMPSTEAD.

BY

HENRY ONDERDONK, Jr.

JAMAICA.
LONG ISLAND
1867.

THIS TRIBUTE TO THE INDUSTRY AND THRIFT

OF THE

ANCIENT SETTLERS OF HEMPSTEAD,

IS RESPECTFULLY DEDICATED TO THE

OFFICERS

OF THE

QUEENS COUNTY AGRICUTURAL SOCIETY.

HISTORY.

Hempstead was settled in 1643, by emigrants from New England, who first bought the land from the Indians, and then obtained a patent for it from the Dutch Governor, the terms of which were: that after ten years from that time, they were to pay the Government a tenth of the revenue arising from the ground manured (*i. e., worked*) by plow or hoe; or, if they should improve their stock by grazing or breeding of cattle, then to make such reasonable satisfaction in butter and cheese as the other towns on Long Island.

This tax in 1657, that is, fourteen years after the settlement of the town, amounted to 100 schepels or Dutch bushels (three pecks each) of wheat. Under the English government this tax was continued by the name of quit-rent at £4 per annum.

The emigrants appear to have settled at first compactly in the village for greater security against Indian hostilities, a fort furnished with commodities for the Indian trade subsequently being built in 1656 at the church, and their flocks and herds driven out in the summer on the great plains to pasture.

The first volume of the town records embracing a period of fifteen years, is unfortunately lost, so that we must ever remain ignorant of much of its earlier history. Though often alarmed with apprehensions of danger, we hear of only one hostile encounter with the natives, and that was caused by stealing pigs. The Indians, though in general inoffensive, would sometimes steal or maim domestic animals, or set their dogs upon them. In this way horses, cows, and hogs were sometimes destroyed. In 1660 the town voted, that no one should sell or give a dog to an Indian under penalty of fifty guilders.

In 1643 there were thirty houses and two hundred or three hundred Indian warriors at Rockaway. In 1671 these had been reduced to ten families, and had forty acres reserved to them for corn. The town, however, forbid anyone to plow or break up any planting land for them, and strange Indians were ordered off. In 1671 the "old Indian wig-wams at Jerusalem" are spoken of.

Though the Indians sold the land, they yet claimed certain rights, such as fishing, hunting, planting corn and cutting basket-wood wherever they could find a suitable tree ; the wood was dyed of various colors and the baskets peddled about the country by squaws.

The first settlers probably found sufficient cleared ground for their purpose. We find in 1708 the barking or girdling of trees on the undivided lands was prohibited under a penalty of six shillings. The "Island of Trees" is first mentioned in 1658. Their houses were constructed of logs, thatched with straw or sedge, and the chimneys built of wooden slats laid in clay. Hence Hempstead, more than once in its early days was endangered by fire, and rewards were given to those who helped to quench it. In 1669 every householder was required to have a sufficient ladder to stand by his chimney under penalty of five shillings ; chimneys were swept and not burnt. (In East Hampton in 1730, the price of sweeping the house chimney was one shilling and six pence ; that of the kitchen nine pence.) On the erection of saw-mills, boards and shingles must have superseded logs and thatch ; the clay-pits furnished brick for chimneys.

Long Island has been called the garden of New York and the crown of the Province ; its fruitfulness has ever been acknowledged. In the Revolutionary war a Tory writer advised the British Minister to land the forces destined for the subjugation of the colonies on Long Island ; "for," said he, "it is 130 miles long and is very fertile, abounding in wheat and every other kind of grain, and has innumerable black cattle, sheep, hogs, &c., so that in this fertile island the army can subsist without any succor from England. It has a fertile plain twenty-four miles long, with a fertile country about it and is twenty miles from New York, and from an encampment on this plain the British army can in five or six days invade any of the colonies at pleasure. The spot I advise you to land at is Cow Bay."

The English did, indeed, land on Long Island, and after the capture of New York, made that city the head-quarters of the army of invasion, and for nearly seven years drew their supplies of fresh and salt hay, oats, straw, wheat, rye, indian-corn, buckwheat and fire-wood from our island, and for an encouragement to farmers to raise plentiful supplies of fresh provisions, vegetables and forage for the army, the British commandant forbid all persons from trespassing or breaking down or destroying fences or carrying away produce from the owners. In 1780 the requisition on Queens County was for 4,500 cords of wood ; in 1782 North Hempstead alone furnished 1,000 cords to the British forces in New York.

CATTLE.

Cattle were imported for breeding as early as 1625, and a cow was worth in New York £30. The abundant grass on the plains doubtless turned the attention of the early settlers to the raising of stock, but as yet there were few or no fences ; so a herdsman was hired by the town to take care of the cattle from the 11th of May till the 23d of October, when the Indian harvest would be wholly taken in and housed. In 1667 the town hired Abm. Smith to keep the cattle from destroying the corn planted or sowed in the plain called the field, and he is to have one and a half bushels per acre paid him for this service. Even at this time complaint is made of birds and worms destroying the corn ; so important was this office (cow-herd) deemed, that the conditions of agreement were entered at large on the town book. At the blowing of a horn, the sun being now half an hour high, the owners of the cattle drove them from their several pens into one common herd, when they were taken under the care of the cow-keeper and his dog and driven on the plains ; he was to keep them from going astray or wandering in the woods or getting on the tilled land ; to water them at some pond at reasonable hours : to drive them weekly to the south meadows ; and then bring them home a half an hour before sunset that they might be milked. For this service (in 1658) the hire was twelve shillings sterling per week in butter, corn and oats.

The number of cattle in Hempstead fifteen years after its first settlement may be inferred from the fact, that seven bulls were kept for the town's use, and that there were then ninety calves that had been weaned and intended to be kept over ; these also at the sound of the horn went out to grass under another keeper on the 2d of June, just a fortnight after their dams had been at pasture. These were to be watered twice a day and taken to the salt meadows once in two weeks and brought home at night and put in an enclosure to protect them from the wolves.

After a while cow-herds were dispensed with, and it was found that fences were necessary for the pasture grounds. Hence we hear (1658) of the East and the West ox pastures. These were enclosed by fences; some of two rails, others of five. Thus Cow Neck (1669) was fenced (as the turn-pike now runs) from Hempstead Harbor to Great Neck, and Rockaway (sometimes called Rockaway Cow Neck) had in 1690 a fence running from the landing across to Jamaica bay. Each proprietor had the right of putting cattle in these pasture grounds in proportion to the length of fence he had made. By degrees the town required the hollows already granted and other cultivated tracts (bevel, tilsome or toilsome and folly— whatever these words mean) to be enclosed against cattle. When clay-pits were imperfectly fenced in cattle sometimes fell in and were drowned.

In 1756 to secure animals grazing on the commons a sure supply of water, highways were laid out to and about several watering places on the plains. In the village there were three ponds, one at the meeting house (Burly pond), one on the east and another on the west end.

After some years a pound for the detention of stray animals was established. In 1708 John Tredwell, Jr., was chosen keeper for the term of seven years, if he behaves as a pounder ought to do and make a good and sufficient pound at his own cost. In 1670 the fine for trespassing on the burial-ground was, for horses and cows, twelve-pence; hogs, six-pence; sheep, four-pence. In 1683 no swine were allowed to go at large after February 1st, unless yoked and ringed. Tame geese were not to run at large (though yoked) on the common after November 5th.

As an instance of the great attention paid to raising cattle, we quote from the inventory of John Smith, Jr., deceased, in 1684. Among the articles enumerated of household goods, are two candle-sticks, seven wooden dishes, ten trenchers, six spoons, and no forks; from the simplicity of his furniture one might reasonably suppose he was in humble circumstances; yet, he was a sturdy, well-to-do farmer, the breeder and owner of at least fourteen oxen, seventeen cows and calves, six steers, two horses and sixteen sheep.

Cattle were sold to the butchers for the New York market, and also exported alive to the West Indies. In 1658 cattle were bought on the great plains of Hempstead, in order to be shipped to the colony of Delaware. In 1678, what is now the city of New York, consumed only four hundred beeves; in 1694 the number arose to near four thousand. In 1682 two oxen were sold in Hempstead at two-pence per pound, and warranted to come to fourteen pound at New York, by weight.

In 1721 a distemper spread among neat cattle, horses, and hogs; and in 1737 Hempstead lost during the winter 850 head of cattle, besides sheep and lambs, for want of fodder.

SHEEP.

Sheep were not introduced in the town so early as cattle. In 1643 there were not over sixteen sheep in the whole colony of New York; they were fed on the great plains (1670) under the care of a shepherd, who had directions not to let them go over half a mile in the woods, for fear of their being lost or destroyed by wolves; no one was allowed to take any even of his own sheep from the common flock or kill it, but in presence of two witnesses; their manure was considered so valuable, that they were folded or penned at night for the sake of their droppings. Cunning farmers sometimes drove by stealth the public sheep and neat cattle into their own private grounds, in order to profit by the droppings; this abuse so increased, that it was deemed necessary by the town in 1726 and again in 1732 to prohibit the folding of sheep or driving them into a close by day or night. As late as 1755, there was a public sheep-pen in the town-spot of Hempstead.

Every proprietor had an ear-mark for his own sheep, which was recorded in the town book; these marks were bought and sold; ingenuity was exhausted in devising new ones. They are described as cropt, slit, nicked, half-penny, slashed, three half-pennies, &c., &c. There were sheep-stealers who altered these marks.

In May, the sheep were parted for washing and shearing. In 1710 the pen was at Isaac Smith's, Herricks; at another time at Success, perhaps for the convenience of having water at hand. After the sheep had fed on the plains during summer, on an appointed day in October or November, the owners, severally, arose early in the morning and commenced driving in the sheep from the outskirts of the plains to a large central pen, then each selected his own by the ear-mark and put them in the smaller pens adjoining. This process was continued till all the sheep were taken out; but if some yet remained without a claimant, they were sold at outcry to the highest bidder and the proceeds went toward paying incidental expenses. The sheep-parting in the fall is of historical interest; it was the great holiday of the times. Here rogues, thieves, and bullies congregated; creditors came in quest of debtors; dealers and traders of all sorts made bargains; horses were swapped, and constables were on the look-out for fugitives from justice; scrub-races, betting, gambling, drinking and fighting, were the order of the day. To counteract these numerous evils, the town enacted a law, that there should be no tavern or selling of liquor at the pens.

HORSES.

The settlers seemed to consider the horse as a beast of drudgery rather than of elegance and speed. True, most of their travelling was of necessity performed on horse-back (sometimes double) through "bridle-ways;" for in a new country wagon paths were not yet laid out. So little regard had they for the comeliness of this animal, that he was subjected to the ignominy of being branded with his owner's name on the buttock and having his ears cropt and slit. Need we wonder then that in 1668 Governor Nichols appointed a horse-race to take place in Hempstead, "not so much," he says, "for the divertisement of youth, as for encouraging the bettering of the breed of horses, which through great neglect has been impaired."

The first course we hear of was on Salisbury plain (so called after Capt. Salisbury) near the Wind-mill pond, now Hyde Park station. This wind-mill was built near the pond, (about 1726) by George Clarke, some time Governor of our State. He called his residence (now Mr. Kelsey's) Hyde Park, after the maiden name of his wife, Hyde. Thence it was removed to the east of the Court House, where it bore the name of New Market till it was removed to the west of Jamaica, and became (1821) the Union Course, where in 1823 an Oyster Bay horse, Eclipse, established his reputation for speed.

ROADS.

In order to illustrate the difficulty of traveling on Long Island in early times before much attention was given to the improvement of roads, we give some "observations" made by Rev. N. Huntting, on his journey from East Hampton to Newtown, at the beginning of the last century. They were noted down in a guide-book that he might not miss his way in traveling.

" Beyond Southampton, about sixteen miles, being about three or four miles from a mill, going over a little brook, just beyond a little wooden causey, and then two paths; leave the right path which goes away to the marsh, and take the left hand path.

" Just over the river by Parker's Fulling-mill leave the right hand beaten road (which goes to Southold) and take a little and blind foot on the left hand.

"A little beyond Coram house leave the right hand path which goes to Setauket, and take a left hand small path by the corner of the field.

" A mile beyond Huntington take the left hand path; about two miles further you come to a new built house and an old one on the left hand, and a mile further take the left hand path.

" Going on to Hempstead plain take the right hand of the two first paths if you would go the back-way and leave Hempstead town; but if you would go through Hempstead, then take the right of the two next paths.

" Going the back-side of Hempstead plain towards Jamaica, being got past Hope Williams' about four miles, entering on another part of the plain, and being come at one house in the corner of a fence with a well before the door, take the left hand path though it be but blind, leaving the plainest path going to houses on the right.

" Going from Jamaica to Newtown, being a little past the last house in Jamaica, take the left hand.

" Going from Newtown to Jamaica, about two miles from Newtown by field, take the right hand path.

" When you come to the first plain past Jamaica houses, if you would go through Hempstead, take Smith Plain path, but if you would go the back-way, take the left hand path.

" Going toward East Hampton, about five miles beyond Huntington by-houses, keep the plain right hand road.

"Going from Lewis's to Coram, just over the river by a field, take the right hand path, the left hand goes to Setaucket.

" Going from Coram toward Parker's mill, take the left hand by the fields, the right hand beaten path goes to the South side of the Island."

Our ancestors, doubtless, undervalued the utility of good roads. In 1675 the town voted ten shillings to clear the way between Hempstead and Little Plains. In 1702 the highway from Jamaica to.New York was so bad as to become the subject of general complaint. In 1808 when a turnpike was projected on this line, the farmers were so opposed to it as to hold an indignation meeting.

HONEY.

Bee-hives are spoken of in 1691, and probably bees were kept long before, as honey supplied the want of sugar. Metheglin and mead with home-brewed beer, cider and domestic wine, gladdened the hearts of our ancestors.

SLAVES.

Slaves were not so abundant in Queens County as in Kings, where a negro with his wife and children occupied the kitchen, which they claimed as their domain; and thus often formed an *imperium in imperio.* They were sometimes lazy and insubordinate. The New Englanders in speaking of a coward fellow, would say: "He is as saucy as a Long Island negro." Being kept from rum, well fed and clad, they were healthy and multiplied exceedingly; so that from 200 blacks in Queens County in 1698, they had grown in 1738, to the number of 1,311. In 1756 the blacks constituted nearly a fourth of the population. In Hempstead eighty-two householders reported a total of 222 slaves, being on an average not quite three to each family; but slavery was not adapted to this part of the Union and was found unprofitable. Emancipation was a boon to the white rather than to the black. The expense of food and clothing often exceeded the value of their labor. It was sportively, but truly said of a farmer who had no corn to sell, "that the hogs had eaten up his corn, and the negroes had eaten up the hogs;" and thus nothing was left at the year's end. After the Revolution, slaves were gradually manumitted, and in 1826 the institution was no more. Jupiter Hammon, a negro slave of Mr.

Lloyd, Queens village, was the author of three publications. Their titles were: 1st. A Winter Piece; 2nd. An Address to the Negroes of the State of New York, 1787; 3d. A second edition of the same, 1806. The horse-rake is said to be the invention of a Hempstead negro. The negroes (bond and free) had a habit of roving from house to house on holidays and Sundays. A mug of cider was accorded them with which they were content; but a dram pleased them more.

In 1683 Thomas Higham sells a slave who has lost the fingers of the right hand and thumb of the left; and in 1687, Christopher Dean sells an Indian boy, slave to Nathaniel Pine. Old newspapers abound in advertisements for runaway slaves. In 1722 Ezekiel Baldwin offers £3 reward for a runaway Indian slave.

TOBACCO.

Tobacco must have been extensively cultivated in Hempstead in early times, as we may infer from the frequent occurrence of such expressions as "weeding tobacco," "stripping off tobacco," "a smoking of it," "planting tobacco on halves," "the old tobacco land," (1676) "a hogshead of tobacco," &c. In 1646 it sold at forty cents per pound in New York. In 1678 John Kissam bought ninety-nine acres of land on Great Neck for £90, to be paid in good merchantable blade tobacco in casks, to be delivered at the weight-house in New York. The culture of tobacco for merchandize gradually fell away; but a little was raised for home consumption, for our frugal ancestors bought nothing from abroad that they could produce at home. The farmers well understood the process of caring of it. After the leaf was stripped off, it was suspended from the rafters of the house to dry. When needed for smoking, it was cut with a knife on a tobacco-board and kept moist in a pouch made of hog's bladder. Chewing-tobacco requiring more skill in preparation, was bought of the manufacturer. In 1737 the farmers got three-pence per pound for their leaf-tobacco.

Raising tobacco has been successful as late as 1833; for the editor of a Brooklyn paper said: "A few days ago we saw two large bales of tobacco on their way to New York. It appeared equal to that of the South, and was raised a little below Hempstead, by an enterprising farmer, from Spanish seed, and was of the 2nd year's planting."

POTATOES.

Potatoes were not mentioned in the early records of Hempstead, and could not have been cultivated till long after the first settlement. A large sort called Bermudian was imported as early as 1636. Our potatoes in olden times were poor, small and watery and quite unpalatable, compared with those now raised. They did not form an important item of diet and were not an article of daily consumption on a dinner table. Perhaps they were as little used as beets, parsnips, or carrots at the present day. Indeed a story is told of a farmer at Flatbush, just before the Revolution, who found his crop amounted to a wagon load; such an excess puzzled his brain. At last he bethought himself to send word to his neighbors to come and carry off as many as they wanted.

TURNIPS.

Turnips were under cultivation before potatoes; but their culture received a great impulse from the zeal and example of Wm. Cobbett, who came to this country in 1817 and took up his residence at Hyde Park, where he raised ruta-bagas in ridge-rows with great success.

GRASSES.

The first settlers did not find the artificial grasses of the old country here, but abundant natural grass on the plains, which made up by its quantity what it lacked in nutrition. The old people say, that the plain grass grew so rank and tall that the dew on it would wet a man's knees as he was taking a morning ride through it on horseback. Poor and coarse as this grass was, it required strict regulations (1697) to preserve its use for the inhabitants. Thus in 1726 an act was passed to prevent firing the grass on the plains, and in 1748 another act forbid the mowing of grass upon the plains before the 28th day of August. It seems that some of the more greedy townsfolk anticipated their neighbors in cutting the grass before it came to full growth. Clover, timothy, lucerne and other grasses have been successively introduced. The advocates of lucerne (1788) claimed that it could be mowed five times in a season and cut eight loads per acre, increased the mess of cow's milk and the quality of the butter, and that it sustained horses as well as oats did, in their hardest labor.

FLAX.

Flax must have been raised from the beginning. Its cultivation involved a vast deal of painful labor, in which the women had their full share. It was pulled by men or boys (not frequently by young women.) In busy times women often lent a helping hand, and usually did the milking at all times) and bound in small sheaves. After harvest a sturdy laborer seizing a sheaf by the butt-end, beat out the seed by striking it on a stone. The seed commanded good prices and was exported to Ireland. The flax-stalk was spread in rows on the grass. When the upper side was sufficiently "rotted," it was turned over with a long pole in order to expose the under side to the action of sun and rain. It was next stowed away in the barn till the leisure hours of winter, when it was set out in the sun and wind, that it might become dry and brittle in order to its being "crackled." After the seed ends had been hatcheled out, it was dressed on a swingle board with a hickory swingle knife. It was then carried into the garret of the dwelling house to be again hatcheled by the women who were kept busy the winter long in spinning the yarn on a foot-wheel. So urgent was this work, that women when invited out to tea sometimes took their wheels with them. Few now remember the little round tea-table in use over fifty years ago, around which about half a dozen ladies sat at a foot's distance, with handkerchief on lap and a tiny tea cup in hand. On the table were a plate of thin-sliced bread thickly buttered, a plate of cake and of smoke-beef, with a saucer of sweet meats, into which each guest dipped her spoon. This yarn was woven (in the family loom frequently) into sheetings, diapers and stuff for domestic wear. Some was knit into stockings, which was a favorite employment or pastime for old ladies and very little girls. The brown linen was bleached white in a "whitening yard" so called. The "tow" resulting from the several hatchelings was spun on a coarser wheel and made into ropes and harness for horses. The summer garments of slaves were woven from tow-yarn. The millenium of women began when cotton, aided by machinery, superseded flax. This was soon after our last war with Great Britain.

WHEAT.

Wheat was probably raised from the first, as the Government tax was paid in that grain in 1657. In 1786 it was so much affected by the Hessian fly, that its cultivation was in a degree abandoned for a time, and rye took its place. Some farmers, I have been told, raised only enough to have flour for cake and pastry and a wheaten loaf on Sundays, and for the regale of visitors. Before the introduction of the German fan and modern threshing machines, the "cleaning up" of wheat and other grain was most tedious and laborious. The grain was threshed out by a flail or trodden out by horses driven in a circle on the barn-floor. Taking advantage of a windy day, the grain was separated from the chaff by tossing both into the air. A hand-fan finished the process. Lewis S. Hewlett had a threshing machine in 1812, but it did not answer any good purpose.

RYE.

Rye was early raised as well for the grain as for the superior quality of the straw for thatching, binding sheaves, &c. When straw was taken to New York for sale, it was unloaded at the York side of the ferry, where it was piled up till a customer came along. The horses were turned out to grass at Brooklyn ferry.

BARLEY.

Barley, though not so often spoken of, was raised to some extent. It was sometimes mowed instead of cradled. It was not usually bound in sheaves, but simply raked up in parcels and thrown upon the wagon by aid of a barley-fork. In 1743 a field of barley was destroyed by caterpillars and worms of an uncommon kind.

OATS.

Oats was always one of the crops.

CORN.

Indian corn was already grown by the Aborigines, who taught the "pale-faces" the virtue of fish as an applied manure.

FERTILIZERS.

The settlers found the soil so fresh and fertile in its natural state, that little or no manure was needed. The hollows, valleys and richer spots of ground were first taken up, but as the soil gradually became exhausted by tillage, home-made manures were used. The animals grazing on the plains were (as far as practicable) penned at night for the sake of their droppings, and the old dead grass was burned that its ashes might fertilize the soil.

In Kings County, manure was imported from New York by (and probably long before) 1768; but it is not known that any was brought thence to Hempstead, till after the Revolution. In 1792 there were manure boats in Cow Bay; and in 1800 (if not before) "spent ashes" were imported from the soap boilers of New York into North Hempstead. After horse-manure rose in price, street "dirt" gradually came in use. It was a long time before the older farmers could realize that high manuring was in reality true economy. No chemist then studied and analyzed our soils and invented appropriate fertilizers. Farmers experimented as well as they could. Plowing in of clover was at one time recommended. Marl, lime, plaster of Paris, seaweed and marine grasses, fish, &c., have been used with various degrees of success.

CURRENCY.

At the time of the settlement of Hempstead, there was little, if any, current coin. The exchanges were made by barter. In bills of sale and contracts, it was specified in what the pay was to consist. It was usually in some of the following articles :

" In good sized wampum ; in good pay ; in corn ; in wampum pay ; for twenty beavers ; in wampum or merchantable goods equivalent ; in corn at current prices ; at beaver prices ; at seventy guilders in good seawant beaver ; in our common pay ; in wampum or corn at current prices ; in New England currency ; in meeting-house pay ; in merchants' pay ; in beaver, or cattle at beaver price ; in corn pay ; at four guilders, or if paid in wampum, then twice that sum ; at prices current equivalent to good merchantable pork, wheat, beaver or seawant."

If the payment was to be made in farm produce, the price per pound or bushel was named in the writing.

In 1673 there was yet but little coin. Wampum was adopted as a circulating medium from the Indians, who manufactured it dextrously with rude instruments, as the heaps of broken sea-shells formerly so abundant along the coast, testify. A hogshead of wampum is spoken of which a man kept in his cellar. This shows how abundant it was.

In process of time copper coin was imported from England, and some silver came in from the Spanish trade and found exports to the West Indies. Portuguese gold coin, such as doubloons, joes and half joes were occasionally met with. Such was the diversity of the coin and some of light weight, that all considerable dealers kept a specie-scales to test the worth of each piece. I have myself seen a gold coin wrapped in a paper, on which was written the goldsmith's certificate of its weight and value.

The first paper money in the colony was in 1709, when our General Assembly issued bills of credit to defray the war debt. These issues were renewed from time to time in order to increase our currency. In 1737, £6,000 was loaned to Queens County. Bills issued in the neighboring States, also found their way on Long Island. In 1723 a school master being detected in passing counterfeit bills, hanged himself in a stable at Hempstead.

At the outbreak of the Revolution, a large issue of paper money was made by the Continental and Colonial Congress, which, however, was soon driven out of our Island by the British silver and gold expended here so freely during the armed occupation.

The last issue of bills of credit by our State, was in 1786. In the course of time banks were chartered by the Legislature.

The settlers used Dutch weight, measures,and money. Hence, we hear of schepels of wheat, ankers of wine, muches of rum, guilders, stuivers, &c.

PRICES.

Our price list cannot be very satisfactory, as we do not know what causes may have influenced prices. No doubt wars abroad and at home, abundance and scarcity of crops, varied the cost of articles from year to year.

BEFORE 1700.

Butter, 6d.; pork, 4½d; beef, 3d. per lb.; wheat per bushel 4/0 to 5/0; corn, 3/0 to 2,6; day's work in harvest, 3/6; hire of a team, 3/6; 100 rails, 5/0; week's board, 5/0; green hides, 3d. per lb.; silver per oz., 6/0; new cart and wheels,

£3 ; catching wolf, 20/0 (in corn) ; Minister's salary, £70 and parsonage, and feed delivered ; Town Clerk, 40/0 per year (in corn); fees of Tax Collector, 8d. on the pound.

ABOUT 1737.

Per pound : beef, 2d. ; butter, 7d. ; leaf tobacco, 3¾d. ; feathers, 22d. ; starch, 8d. ; bacon, 4½d. ; sugar, 8d. ; linen yarn, 2/10.

Per bushel : wheat, 3 3 to 4 0 ; rye, 2/6 ; oats, 1/6 ; corn, 2 8.

Per gallon : molasses, 21d. ; rum, 3/6 to 5/0 ; vinegar, 6d. to 9d. ; milk, 6d.

Day's wages of negro, 2 3 ; of a washer woman, 1/3.

Turkeys, 5 0 each ; chickens, 5d. ; pigs, 3/0 ; calf-skin, 2 6 ; hind quarter of very good veal, 2 6 ; of mutton, 2/9 ; of lamb, 1/9 ; 60 eggs, 1/6 ; a Dutch plow, £1 3/3 ; a wagon and tackling, £4 10/0 ; scythe and tackling, 7/4 ; 100 rails, 12 0 ; midwife's fees, 14/0 ; fourteen pounds of butter was exchanged for half a pound of tea ; barrel cider, 6/0.

1750 TO 1760.

Per pound : 2¼d. to 2½d. for beef ; pork, 4¼d. ; gammon, 6d. ; wool, 1/3 ; mutton 3d. to 4d. ; veal, 3½d. ; leather, 1s. 2d. ; candles, 9d. ; flax, 7d. ; butter, 1 0 ; cheese, 6d. ; chocolate, 2 6 ; soap, 1s 2d ; iron, 4d. ; tobacco, 5d. ; lath nails, 1 3 ;

Per bushel : wheat, 5/6 ; rye, 4/6 ; corn, 3/0 ; flax-seed, 3 6, 5/0 and 10s. ; clams, 6d. ; potatoes, 1s. 6d to 3s. ; oats, 2s. ; bran, 10d. ; turnips, 16d. ; lime, 12d.

Per gallon : tar, 2s. ; milk, 1s. ; vinegar, 1s. ; rum, 4s. ; wine, 12s.

Day's work mowing, 3s. 3d. ; ordinary work, 2s. 6d. ; young apple trees, 6d. ; marriage fees, 8s. to 24s. ; day's work carpenter, 5s. ; digging a grave, 6s. ; beaver hat, £2 8s. ; a wig, £2 16s. ; a pewter spoon, 6d. ; a candlestick, 1s. 3d. ; a broom, 5d. ; tea-kettle, 20s. ; a dozen knives and forks, 6s. 6d. ; ox-load of walnut wood, 3s. ; a christening, 20s. ; English hay per cwt., 5s.

PHYSICIAN'S FEES, 1730.

An emetic, 1s. 6d. : bleeding and attendance, 2s. ; drawing a tooth, 1s. ; journey at night and attendance, 1s. 6d. ; two cathartics, 3s. ; two paregoric draughts, 5s. ; anodyne pill and journey, 2s. ; eye-water, 2s. ; two doses Ethiop's mineralis, 2s. ; innoculating four children, 4s. ; opening an imposthume, 1s.

1740 TO 1750.

Bleeding, 1s. ; emetic, 1s. ; *linctus ad tussem*, 9d. ; styptic, 1s. ; sudorific, 1s. ; visit 3s. ; visit in the night, 4s. ; electuary, 5s. ; jalap, 2s. ; opening a tumor, 1s

6d. ; three blister plasters, 1s.; potion powder for diarrhœa, 1s.; large anodyne, plaster, 2s. 6d. febrifuge decoction for diarrhœa, 8s. ; seven febrifuge pills, 2s. 4d. ; one visit and writing patient's will, 3s. ; visit and advice to a child, 1s. ; *ad aperiendum tumorem*, 1s.

Our ancestors had great faith in native herbs and roots. The list would more than fill this page. Clergymen were sometimes doctors; also as was the Rev. Samuel Seabury, of Hempstead; but they all kept a book of recips in which the virtues of roots and herbs were duly set forth ; and thus were often consulted for advice by their sick parishioners. Dr. Searing was a most noted *worm* doctor. He was also famed throughout the country for curing *jaundice.*

There were strange diseases then as now that overspread the country and baffled the skill (such as it was) of the physicians; such as small-pox (1730) ; malignant sore throat distemper (1769) ; great sickness (1668); pleurisy (1728) ; dysentery (1790) ; intermittent, remittent and billous fevers in North Hempstead in 1820, 1829, 1830 and 1831.

FISHERIES.

It would be hardly possible to exaggerate the value of the fisheries of Long Island. Great numbers of persons on the north and south shores have, time out of mind, pursued a lucrative employment in catching fish, eels, oysters, clams, scollops, crabs, &c., for the New York markets ; but the statistics are to us unknown. As early as 1643 the Indians at Rockaway feasted their white visitors with oysters and fish ; and as late as 1667 they dried oysters and clams for a tribute to their more warlike brethren on the Main.

The whaling business must have been followed to some extent by the people of this town, for in 1710 the town granted the " woodland at Rockaway for the use of the whalemen of Hempstead to cut firewood from for the use of the whaling design." Whale neck is spoken of in the records in 1658 and 1684.

To such an extent did interlopers rake up the clams and break their shells to make lime of, that it was forbidden by a town act in 1753. The value of horse-feet for chickens and swine is well known.

TREES.

Fruit trees have been more successfully cultivated on the north than on the south side of our island. Apple, pear, peach and cherry trees were probably nearly coeval with the settlement. An orchard is spoken of in Hempstead in 1657. The peach was proverbially abundant. The trees were overloaded and the fruit shaken off daily for the swine. Several exotic fruit trees have been introduced, such as the Madeira nut, but our severe winters bear hard upon them.

Ornamental trees have attracted the attention of the more enlightened farmers and there have been several tree-fevers. The Lombardy poplar was introduced (1790) and had a great run. Its virtues were grossly exaggerated, its leaves were claimed to be better fodder for cattle than hay. Long, stately rows by fences and highways were to be seen fifty years ago. The white mulberry, (Mr. Aspinwall, of Flushing had a nursery of white mulberry trees in 1760) whose leaves were the food of silk worms, also had its day. Who does not recall to mind the morus-multicaulis and the ailanthus? A Virginia trader is said to have planted the first locust at Sand's Point, whence it has spread over the Island.

When we have such a variety of nuts at home, why should we send our money abroad for those but little (if at all) superior to our chestnuts, butternuts, black walnuts and hickory nuts?

The father of the writer (whose zeal for propagating trees is well known) planted a nut from which grew a black walnut tree of such dimensions, that when sawn up into plank, it furnished coffins for himself and wife.

WOOLLEN MANUFACTURES.

From the wool shorn from the sheep the women carded rolls. These, by a large woollen wheel were spun into yarn, which was woven in the domestic loom. This cloth was, perhaps, at first worn without being colored, fulled or dressed; but in time, carding, fulling and shearing mills were set up on sites granted by the town on numerous streams. Leather cloths of sheepskin were much worn by the laboring class. Buckskin formed part of a tailor's stock in trade. In 1775 Congress recommended to farmers and others not to kill lambs, but to preserve them for the sake of their wool. Shoes were sometimes made of dog's skins.

INVENTIONS.

Patents have been granted to the inventive genius of Queens County, as follows:

1814 to Peter Baker, Long Island, machine for cutting grass.
1821 " Andrew Cooke, Flushing, seeding corn planter.
1821 " Thomas Morgan, Hempstead, manufacturing nails.
1832 " Homer Whittemore, Newtown, cards, boards, &c.
1812 " Rev. Seth Hart, Hempstead, cloth dressing.
1811 " Jessie Molleneux, " shearing cloth.
1836 " Rev. W. H. Carmichael, D. D., Hempstead, stoves.
1820 " Wm. R. Loweree, Flushing, machine for propelling boats.
1814 " David Cooper, Jericho, water wheel.
1817 " Jarvis Smith, Queens County, a flutter letting water on a water wheel.
1823 " Jessie Mollineux, North Hempstead, a wind wheel.
1811 " Daniel Voorhies, Long Island, a wind wheel or water wheel.
1824 " Nathaniel W. Conklin, Jamaica, a saddle spring.
1815 " Peter Cooper, Hempstead, a cradle.
1819 " Jeremiah H. Pierson and J. H. Simmons, Hempstead, a loom.
1813 " John J. Staples, Flushing, wind drawing machine.
1799 " Rev. Seth Hart, machine for manufacturing nails.

Rev. Mr. Hart, Rector of the Episcopal Church, Hempstead, was of an inventive turn of mind. In addition to his ministerial duties, he kept a classical boarding and day school. He was at great expense in a fruitless attempt to secure a patent in England for his cloth dressing machine. He also failed in the manufacture of brooms. So many avocations led to a neglect of his flock, and many passed into the Methodist fold.

MECHANICS.

Mechanics were always acceptable in a new settlement. In Hempstead, as an inducement for this useful class to come and settle business among them, grants of land were freely made to the cooper, the blacksmith, the miller, and the like. In 1691 John Stuard petitioned for a grant of land, wishing to establish himself as cooper or surgeon, though what is meant by " surgeon" we cannot say.

MANUFACTORIES.

There were sundry manufactures that flourished for a while, such as a pot-ash factory at Herricks, owned by Jos. Burr, 1773. I have seen an earthen sugar cup made at a pottery on Cow Neck. Thomas Parmyter had extensive works at Whitestone (1736) for making tobacco pipes and flower-pots. In 1728 Josiah Millikin, at Glen Cove, made periwigs. James Mott, of Westbury, (1816) was an ingenious weaver of diapers, coverlets, &c. Alfred Hentz, (at Greenwich) clarified quills, which he gathered in his own wagon about the country. He also gave lessons in French. A paper mill was set up at Hempstead Harbor, by Hendrick Onderdonk, in 1773. Buttons of apple tree wood were made by Samuel Wood, at Searingtown, at the beginning of this century; they had a metal shank. Owing to the impulse of the war of 1812, a cotton factory was erected at Manhasset, and the Nassau Woollen Factory at Roslyn. The Flushing Manufacturing Company was organized in 1813. Who has not heard of the woollen manufactures of John H. and Walter R. Jones, of Cold Springs?

SPORTS.

Our ancestors, though industrious and frugal, yet knew that "all work and no play makes Jack a dull boy." Accordingly, they had their relaxations from toil; such as hunting, fowling, going in the bay, fishing for shell and scale fish, horse-racing, huckle berry frolics. helping spells, raisings, sheep partings, attending courts and vendues, looking up meadow hens eggs, beach parties, &c. (A huckle berry frolic was, originally, a party of young folks going to the woods to pick berries; they took a lunch with them. When they returned they often stopped at a tavern, and with the aid of a fiddler finished the day with a dance. Now, it consists chiefly in a series of scrub-races and riotous mirth. A beach-party, in which young men and women take a ride to the Rockaway beach, is sometimes called a huckle-berry party, because it took place at that season of the year.)

At funerals they had a good time of it. Custom and respect for the memory of the deceased required abundant refreshments. At the funeral of the minister of Hempstead, in 1764, the sum of £3 10s. was expended for wine; five gallons of rum were bought for the vendue of his effects; at a funeral of a child in Rev. Thomas Payer's family, (1730) Jamaica, five gallons of rum were used; in Nicholas

Tanner's will (1658) he orders a cow to be killed at his burial and given to his neighbors.

They had home-made beer and wines; also French and Spanish wines found their way into Hempstead before 1658. If the town made a grant of land or other privilege to anyone, he was expected to treat. Thus, when John Ellison (1676) had a grant of four acres at Great Neck, he gave the voters two gallons of rum to drink. Thomas Rushmore (1683) had a silver dram-cup.

FAIRS.

The fairs spoken of in 1728 and 1774 were rather market days for the sale of merchandize, stock and farm produce. It was in 1819 that the first fair was held in Queens County. This was discontinued at the end of four years for lack of encouragement. It was in 1842 that the present Society held its first fair at Hempstead, which has steadily increased in interest, and its success has culminated in the establishment of its noble permanent ground and appurtenances at Mineola.

CONCLUSION.

The thrift of our early settlers was due to their industry and frugality. Chocolate, tea and coffee had not yet taxed their resources; milk, with bread and butter, hasty-pudding and homony constituted the principal elements of the morning and evening meal. Spoons (of horn or pewter) they had from the first, but not forks. Cider, bread and meat, and a few vegetables (but not potatoes) were on their dinner table.

One or (at most) two candlesticks sufficed for a family that kept early hours. Even more than a century later, 365 candles lasted a year in a farmer's family. A boy or girl was expected to make one pair of stout shoes last the year round.

Fashions did not change, so that the same dress might descend from parent to child till fairly worn out; they bought nothing that they could make at home.

Samp was made from whole grains of corn hulled in a large wooden mortar with a heavy pestle. This was a negro's task of a winter evening.

Almost every farmer's son learned some trade—not that he must necessarily follow it ; but that he might have something to rely on for a living.

Education was not neglected by the settlers ; for we find that the town (in 1658) devoted the fines from unlicensed dram-sellers to the schooling of poor orphans. In 1662, Jonas Houldsworth was schoolmaster, then Richard Gildersleeve, who gave place to Richard Charlton in 1670. In 1702 the town voted 100 acres of land for a free school ; also timber for building, fencing and firewood. (Rev. Samuel Seabury, who died in Hempstead in 1764, taught a classical school. Next came Rev. L. Cutting, 1766; then Rev. Thomas Lambert Moore, 1785 ; Rev. Seth Hart, 1802 ; Rev. Timothy Clowes, L.L.D., 1819 ; all these clergymen taught classical and boarding schools of high reputation) In 1707 four acres of land, west of the meeting house pond, was granted to "settle a school master upon for to teach our children—the land to be for the use and privilege of a school for ever."